WITHDRAWN

W9-ANX-480

Writing and Numbers

Written by
Nigel Nelson

Illustrated by
Tony De Saulles

Thomson Learning

New York

Books in the series
Body Talk
Codes
Signs and Symbols
Writing and Numbers

Picture acknowledgments
Ancient Art and Architecture Collection 5, 10 (top), 14 (bottom), 18 (both); Archiv
Für Kunst und Geschichte 4 (top), 6, 8; Daniel Pangbourne *cover*, 9, 17 (top), 22,
26; C. M. Dixon Photoresources 14 (top); 17 (right); Eye Ubiquitous 20, 25
(bottom); Eye Ubiquitous/Frank Leather 12; Mary Evans Picture Library 17 (left);
Tony Stone Worldwide/Jean-Pierre Gerard 10 (bottom); Wayland Picture Library
13, 19 (top), 27; Werner Forman Archive 4 (bottom), 11, 16; Tony Woodcock
Photolibrary 29; Zefa 19 (bottom), 21 (top), 25 (top).

First published in the
United States by
Thomson Learning
115 Fifth Avenue
New York, NY 10003

First published in 1993 by
Wayland (Publishers) Ltd.

Library of Congress Cataloging-in-Publication Data applied for

ISBN 1-56847-158-0

Printed in Italy

Contents

Words that are printed in **bold** are explained in the glossary.

Cave painting

The first people on earth did not know how to write. All they could do was paint pictures on the walls of caves and on rocks, some of which can still be seen today. This was the earliest form of writing. Writing is a way of communicating by making marks or signs that stand for the spoken words of a language.

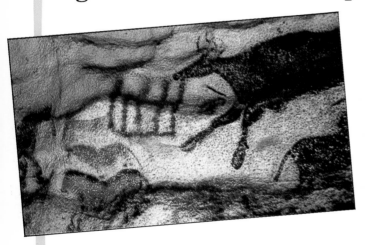

These little horses and the jumping reindeer were painted in a cave in France thousands of years ago.

This drawing of an elephant and hunter was found on a rock in Africa. Pictures like this tell us what kinds of animals people hunted long ago.

Native American people told stories with pictures of people and animals. These figures were drawn by Native Americans living in what is now Utah.

Some cave paintings had signs that we do not understand. Cave pictures and signs were probably the first steps towards writing.

Activity

Make your own cave painting on paper. Use charcoal and pastels. Cave artists used only a few colors, usually black, brown, yellow, and red.

Writing in pictures

The first known system of writing was invented over 5,000 years ago by the **Sumerians**, who lived in what is now southern Iraq. The Sumerians used pictures to stand for words.

These Sumerian picture-words looked very much like the things they stood for. For example, the Sumerian picture-word for a bird looked like a bird.

The **ancient Egyptians** got the idea of picture-writing from the Sumerians. In another part of the world, the Chinese also developed a form of picture-writing about 4,000 years ago.

This map shows you where the Sumerians, Egyptians, and Chinese lived.

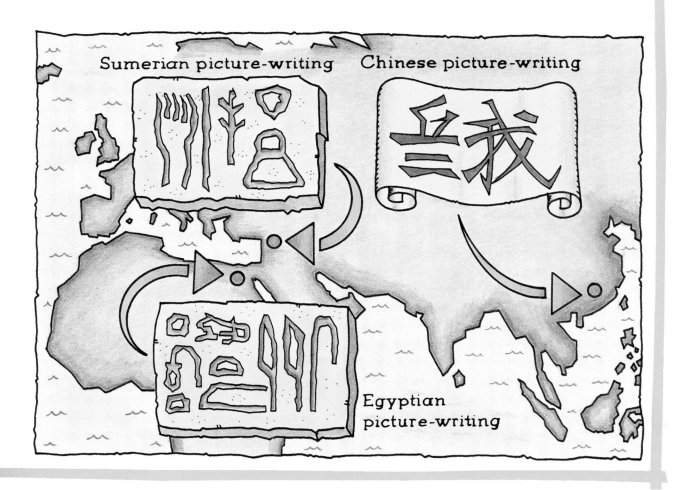

Sumerian picture-writing

Chinese picture-writing

Egyptian picture-writing

Sumerian writing

The first Sumerian picture-words were carved into stone or metal. Later the Sumerians wrote by pressing their words into slabs of wet clay, using a flat stick. The clay was then put out in the hot sun to bake hard. It is easier to draw straight lines in clay than it is to draw curves, so the picture-words had straight lines. As you can see from this picture, Sumerian clay writing was wedge-shaped. This style of writing is called cuneiform, which means wedge-shaped.

Sumerian cuneiform words were written with fewer marks than the first picture-words, so they no longer looked like the things they stood for. Cuneiform writing was a sort of code that had to be understood by both the person who wrote the words and the person who read them.

Activity

Make up your own cuneiform-style writing. Press the words into a slab of modeling clay using a flat stick. Is it easier to draw straight lines or curved shapes?

Egyptian writing

The ancient Egyptians copied the Sumerian way of writing.
Egyptian picture-words are known as hieroglyphs. It takes many years of study to read Egyptian hieroglyphs like these.

The first hieroglyphs were carved into stone, like the ones carved on the entrance to this **temple** at Abu Simbel.

Later, the Egyptians made a paper-like material from **papyrus**, which is a plant that grows along the Nile River. They wrote on the papyrus with pens made from **reeds** and ink made from soot, water, and **gum**. As time went on, the Egyptians wrote their picture-words with fewer lines. Eventually the hieroglyphs were written as simple shapes.

These hieroglyphs written on papyrus are much simpler than hieroglyphs carved into temple walls.

Chinese writing

The Chinese also developed a form of picture-writing. Over the years the pictures were changed into the shapes that are still used by Chinese people today. The picture-words are called ideograms. These posters are written in Chinese ideograms.

Each ideogram stands for a word in the Chinese language, but there is no way of knowing how to say the word.

People in different parts of China speak different forms of Chinese. They may often use different sounds for a certain ideogram, but everyone knows what the ideogram stands for. So, even if two Chinese people cannot understand each other's spoken words, they can read the same books and newspapers and even write letters to each other.

The Chinese were the first people to make paper from wood **pulp**. They wrote with ink and brushes and took great care to make their writing look beautiful.

Alphabets

An alphabet is a set of letters that stand for sounds and not whole words. The letters of an alphabet can be arranged to form all the words in a language.

The first alphabet was invented by the **Phoenicians** about 3,000 years ago. The ancient **Greeks** based their alphabet on the Phoenician alphabet. The words carved into this stone are written in the Greek alphabet.

These **Latin** words were carved into stone by the ancient **Romans.**

The Romans used some of the Greek letters to develop their own alphabet. The Roman alphabet is the one used in most European languages today.

This map shows how the use of an alphabet spread from the Phoenicians, to the Greeks, and then to the Romans, and finally to the rest of Europe and other parts of the world.

How our alphabet came to us

North America

Europe

Rome

Greece

Phoenicia

Australia / New Zealand

Books

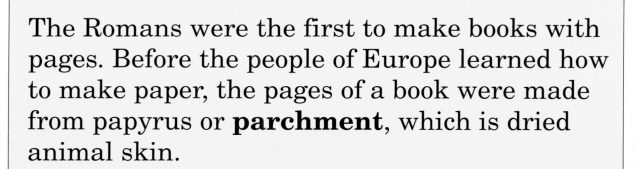

The Egyptians invented the book. They pasted sheets of papyrus together to form one long sheet. This was rolled up like a roll of wallpaper to make a scroll. This carving shows an Egyptian writing on a papyrus scroll.

The Romans were the first to make books with pages. Before the people of Europe learned how to make paper, the pages of a book were made from papyrus or **parchment**, which is dried animal skin.

Long ago, the only way to make a copy of a book was to write it out by hand. In Europe, **monks** used to copy books with **quill pens**.

Activity

Draw and paint your
own illuminated letters.
Use gold and silver
paint if you have some.

The monks often drew beautiful decorations on
the first letter of a page. These decorations are
called illuminations.

Printing

This is the first known printed book. It was printed in China over a thousand years ago.

The ideograms that made up a section of the book were carved into a block of wood, like the one shown here. The block was covered with ink, then the paper was pressed on top.

Over 500 years ago, a German named Johann Gutenberg invented a new way of printing using metal letters. The letters were arranged in a wooden frame to make up a page of a book.

The letters were then covered with ink and pressed against paper using a screw press like the one shown here. Once all the copies of a page had been printed, the type could be rearranged to form another page of the book.

With the invention of the printing press, books could be copied much more quickly. Today, printing is done with modern presses and computers. More books, magazines, and newspapers are printed than ever before.

Writing for the blind

About 100 years ago, a Frenchman named Louis Braille invented an alphabet that could be read by blind people. In the Braille alphabet, each letter is made up of a group of raised dots, or bumps. A blind person can read the Braille writing by feeling the raised dots on the page.

The bumps on this watch help the wearer tell time by touch.

This girl is reading a book by touching the raised dots with her fingertips.

Activity

These are the letters of the Braille alphabet.

A	B	C	D	E	F	G	H	I	J

K	L	M	N	O	P	Q	R	S	T

U	V	X	Y	Z	and	for	of	the	with

Write a message using the Braille alphabet. Make the raised dots by pressing a sharp pencil into the back of a piece of cardboard.

Numbers and counting

The first people probably used their fingers to count. Like the boy in this picture, they held up their fingers to show which number they meant. But we have only ten fingers, so people needed some way to count past ten.

One way of counting past ten was to make marks or scratches in the ground or on stone. Over time these marks developed into **symbols**, or signs, that stood for numbers. Symbols that stand for numbers are called numerals.

Here are the numerals used by the ancient
Egyptians. The Egyptians used simple marks
arranged in rows for the first nine numbers.
Their numeral for the number ten was a
horseshoe shape.

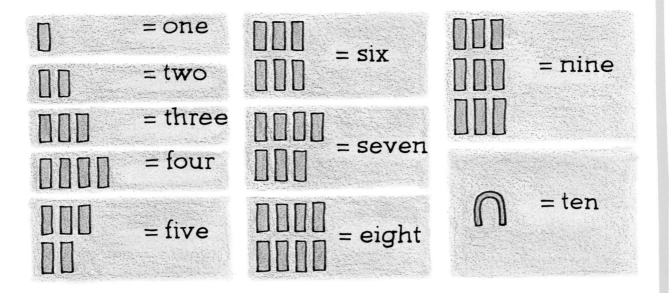

Here are some of the other Egyptian numerals.

Roman numerals

About 2,000 years ago the Romans started to use letters from their alphabet to stand for numbers. The Roman numerals from one to ten are shown here.

I	=	one
II	=	two
III	=	three
IV	=	four
V	=	five
VI	=	six
VII	=	seven
VIII	=	eight
IX	=	nine
X	=	ten

The Roman numerals for eleven to twenty are made by adding X (10) in front of the first ten numerals. So, the Roman numeral for twelve (10+2) is XII, and for eighteen (10+8) it is XVIII.

The Romans used other letters for larger numbers: L = 50, C = 100, D = 500, and M = 1,000.

Romans numerals are sometimes used today. The numbers on this clock are written in Roman numerals.

The date on this building is written in Roman numerals. The words are written in Greek.

The abacus

The main problem with Roman numerals is that they are difficult to add up. Try doing this sum in Roman numerals: IV + XVIII. The answer is XXII. But to get the sum you had to add the figures up in your head because the numerals do not add up easily in columns. To help do addition, the Romans used an abacus. You may have an abacus in your school like the one this girl is using.

In the abacus the Romans used, the row of beads on top stands for units, or ones. The next row stands for tens; the next for hundreds; the next for thousands; and so on.

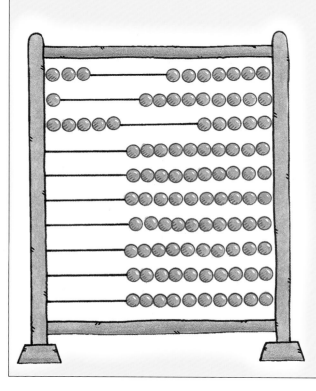

To show the number 3, you move the first three beads on the units, or ones, row. To show 13 (3 + 10), push one bead on the tens row and three on the units row. To show the number 513 (500 + 10 + 3), push five beads on the hundreds row, one on the tens row, and three on the units row.

In some countries, people still use the abacus. First the highest digit is set up, then the next highest, and so on. The abacus beads show the sum.

Arabic numerals

0 1 2 3 4 5 6 7 8 9

Most people today use **Arabic** numerals. These numerals were developed in India but were also used by the Arabs. The people of Europe learned them from the Arabs so they called them Arabic numerals. The ten Arabic numerals are shown above. Each numeral is called a digit. Using these ten digits we are able to write all our numbers.

The value of a digit depends on its place, or position, in a number. A digit to the left always counts in numbers that are ten times bigger.

This chart shows how to work out the value of each digit in a number.

thousands	hundreds	tens	ones
		6	2
	8	6	2
1	8	6	2

In the number 62, 6 means six tens (6 × 10) and 2 means two ones (2 × 1).
In the number 862, 8 means eight hundreds (8 × 100).
In the number 1,862, 1 means one thousand (1 × 1,000).

With the invention of Arabic numerals, people no longer needed an abacus to do their sums. Numbers could now be added up easily in columns.

Glossary

Ancient Belonging to times long past.

Arabic Anything connected with the Arabs, who are the people living in Arabia and other parts of the Middle East.

Egyptians People who live in Egypt, a country in northeast Africa. The history of the ancient Egyptians goes back about 5,000 years.

Greeks The people who live in Greece.

Gum A sticky material that is found in some trees and plants.

Latin The language of the ancient Romans.

Monks Men who live apart from the rest of society to worship God.

Papyrus A tall, grassy plant that grows along the shores of the Nile River in Egypt. Our word "paper" comes from "papyrus."

Parchment The skin of an animal, usually a sheep or a goat, that has been dried and stretched. Parchment was used in the past as a sort of paper.

Phoenicians An ancient people who once lived in the Middle East.

Pulp Wood that has been ground up and moistened so that it can be made into sheets of paper.

Quill pen A simple pen that was made from a bird's feather.

Reeds Plants with tall stalks that grow in or near water.

Romans People from Rome, in Italy. About 2,000 years ago, the ancient Romans ruled over most of Europe and parts of northern Africa and the Middle East.

Sumerians An ancient people who used to live in a region called Sumer, which is now part of Iraq.

Symbol A mark or sign that stands for a thing or an idea.

Temple A building that is used to worship a god or gods.

Books to read

Carona, Philip. *Numbers.* New True Books. Chicago: Childrens Press, 1982.

Elliott, Deborah. *Making a Book.* New York: Thomson Learning, 1993.

Cobb, Vicki. *Writing it Down.* New York: HarperCollins Childrens Books, 1989.

Index